ZOOM!

THE

BUGS

Camilla de la Bédoyère

NEW BURLINGTON BOOKS

A NEW BURLINGTON BOOK
The Old Brewery
6 Blundell Street
London N7 9BH

Editor: Amanda Askew
Designer: Andrew Crowson
Picture Researcher: Maria Joannou

First published in the UK in 2011 by
QED Publishing
A Quarto Group company
226 City Road
London EC1V 2TT

www.qed-publishing.co.uk

A catalogue record for this book is available from the British Library.

ISBN 978 1 78171 369 3

Printed in China

Picture credits
DK Images 19tr, 32;
FLPA Mark Moffett/Minden Pictures 6b, Rolf Nussbaumer/Imagebroker 7bl, Thomas Marent/Minden Pictures 10b, Cisca Castelijns/FN/Minden Pictures 11cr, Mark Moffett/ Minden Pictures 12b, Piotr Naskrecki/Minden Pictures 13b (background), Mark Moffett/ Minden Pictures 14b, Murray Cooper/Minden Pictures 15b, Ingo Arndt 16b, Chien Lee/ Minden Pictures 16t, Murray Cooper/Minden Pictures 17tl, Ingo Arndt 18b, Frans Lanting 19b, Michael Durham/Minden Pictures 20t, Mark Moffett/Minden Pictures 21b, Steve Trewhella 21cr, Albert Lleal/Minden Pictures 21tl, John Eveson 22-23t, Heidi & Hans-Juergen Koch/Minden Pictures 22br, Mark Moffett/Minden Pictures 23b, Gary K Smith 23t, James Christensen/Minden Pictures 24b, Albert Mans/FN/Minden 24t, Gerry Ellis/Minden Pictures26-27, David Hosking 27bl, Mark Moffett/Minden Pictures 28cl, John Eveson 28br, Albert Mans/FN/Minden 29cl, Albert Lleal/Minden Pictures 29tl;
Nature Picture Library Nature Production 5t, Stephen Dalton 8b(2), Doug Wechsler 17b, Nature Production 21tr, Meul/ARCO 22bl, Nick Upton 25br, 27br, Stephen Dalton 27t, Doug Wechsler 29cr;
Photolibrary Flirt Collection/LWA-Dann Tardif 4t, Eastphoto Eastphoto 9tl, Age Fotostock 12t, Emanuele Biggi 13tl, Antonio Lopez Roman 14c, Last Refuge 20b, Martin Ruegner 25bl, Otto Hahn 25t, Age Fotostock 28bl, Otto Hahn 29bl;
Photoshot NHPA/Mark Bowler 7t, NHPA/Stephen Dalton 13b, 17tr;
Science Photo Library Tom McHugh 4b, Richard R Hansen 5b, 7br, Claude Nuridsany & Marie Perennou 8b(1), 8b(3), 8b(4), Tom McHugh 13tr, Eye of Science 18t, Tom McHugh 28cr;
Shutterstock Karbunar 6-7c, Studiotouch 8t, Pan Xunbin 9b, Eric Isselée 9cr, Dr. Morley Read 9tr, Chas 10-11, Pan Xunbin 11b, Alslutsky 11t, Liew Weng Keong 14-15t, Dr. Morley Read 15tr, Tischenko Irina 18-19, Matthew Cole 26b, Tischenko Irina 28t, Chas 29br, Eric Isselée 29tr, Matthew Cole 30
Front and back cover Matt Cole

Words in **bold** can be found in the Glossary on page 30.

▼ Flies have a short, streamlined body. They have compound eyes on the side of their head and small antennae. They only eat liquid food.

CONTENTS

ZOOM INTO...

...the world of bugs and begin a journey that takes you to a hidden place, full of mystery and surprises. You could hold a bug in your hand, but how much would you really know, or understand about it? Now you can imagine what it is like to be bug-sized, and lead a life filled with excitement and adventure. Get ready to discover some amazing facts and become closer than close to bugs with amazing ZOOMs!

Zoom in

Look at an ant through a microscope, or through a **macro lens** on a camera, and you will notice that the ant appears bigger than in real life. A microscope uses lenses to magnify the image of very small things, often several hundred times. Modern microscopes may use other techniques to magnify objects many thousands of times.

Almost real

Bugs with the ACTUAL SIZE icons are shown at their real-life size, as though they're crawling across the page! Comparing the bug to a standard paperclip really helps you to understand its size.

Macro photography

The art of taking pictures of small things in close-up is called macro photography. Using these, and other techniques, photographers and scientists have helped us to uncover a world we never knew existed, by giving us a bug's eye view. Images with the ZOOM icon show you how many times these bugs have been magnified.

Try it

WHAT IS IT? images let you use your new investigation skills to guess what the bug might be. Then just turn over the page to find out that IT IS...

WHAT is it?

READY FOR ACTION

Most bugs are small enough to hide easily. So if your survival depended on finding and catching thousands of them, you would want to be well equipped. **Mantids** have it all – great eyesight, lightning reactions and legs that are covered in gripping spines.

Wait and see

Most mantids are green, so they can hide from their prey among leaves. They sit absolutely still, but they are alert and ready for action. When a mantid sees lunch approaching, it prepares for attack, and succeeds in less than one thousandth of a second – that is about 300 times faster than the blink of an eye.

Praying mantid

While the front half of the body lunges forwards, the back half stays still.

Posed for action

The mantid's front legs are bent, and held in a strange pose that makes it look as if it is praying. In reality, these spiked legs are ready for a much less peaceful activity. Fast and strong, the vice-like limbs can catch and crush an insect in less than a second.

FACTOID

Some mantids are big enough to hunt for larger prey. They can attack birds and lizards.

VITAL STATISTICS

Common name	Praying mantid
Latin name	*Mantidae*
Size	12–150 mm in length
Habitat	Forests
Special feature	Lightning-quick predator

It is...

a giant leaf insect, which has camouflaged itself by pretending to be part of a plant. When an animal pretends to be something else, it is called a mimic. Look closely, and you will see this bug has fake 'veins' and brown edges, just like a real leaf.

GROWING UP

Look deep into the world of bugs and you will find incredible changes occur as their lifecycles unfold. Take an ugly blowfly as an example. With its big eyes, bristly body and shiny green **cuticle**, this fly is easy to recognize. But would you recognize its young so easily?

◀ Blowflies with a green metallic sheen are called greenbottles.

All change

*Most bugs begin life looking completely different to the way they will look as adults. They go through extraordinary changes as they grow. Each change is a **metamorphosis** – which means 'transform shape'.*

The blowfly's lifecycle

1. *A female blowfly can smell a dead body, open wound or meat. Within minutes, she arrives to lay about 250 eggs on it. Just 24 hours later, the eggs hatch into tiny **grubs**, called maggots or **larvae**.*
2. *The maggots feed on the flesh or food and as they grow, they **moult**. After two moults, the maggots are much bigger and will soon be ready to metamorphose.*
3. *Each maggot grows a tough, dark case around its body and becomes a **pupa**.*
4. *It takes at least six days for the pupa to transform into an adult. When the pupa breaks open, an adult blowfly emerges.*

Silky case

Some bugs wrap themselves in a silk ***cocoon*** *when they pupate. They make the silk in special silk glands in their bodies. The larvae of silkworm moths are kept in large farms, and the silk from their cocoons is used to weave silk for clothing.*

Stripping off

When a bug grows, it gets too big for its tough skin, or cuticle. The simple solution is to strip off the old skin – this is called moulting – to reveal a new, bigger cuticle underneath.

Giant grubs

The larva of a Hercules beetle can reach up to 16 centimetres long and weigh 150 grams, making it one of the largest grubs in the world. It feeds on soft, rotting wood.

WHAT IS IT?

BIG-EYED BUGS

How great would your eyesight be if your eyes covered your whole head? Members of the dragonfly family are lucky enough to have enormous eyes and great vision. Their eyesight is so good, these insects can see in almost every direction. They can see colours (except red) and even **ultraviolet** light, which is a type of light that humans cannot see.

A trick of the light

The Southern Hawker dragonfly has 30,000 ***lenses*** *in each eye. A lens focuses light that enters the eyes, and nerves carry the image to the brain. There, the brain takes all 60,000 images and turns them into a fantastic ability to see. Insect eyes are especially good at detecting movement. Human eyes have just one lens each – but the quality of images that our brains can see is much better.*

Dragonfly

Air acrobatics

Members of the dragonfly family aren't just superb at seeing, they are also among the world's best fliers. They can twist and turn each of their long, lacy wings independently, so they can fly forwards, backwards, change direction quickly and even hover.

FACTOID

Prehistoric dragonflies were enormous, with a wingspan of 75 centimetres!

VITAL STATISTICS

Common name	Southern Hawker dragonfly
Latin name	*Aeshna cyanea*
Size	70 mm in length
Habitat	Near ponds and rivers, in gardens and woodlands
Special feature	Superb flier

It is...

the larva of a dragonfly. These youngsters are called nymphs (say: nimfs) and they live in water, where they hunt worms, tadpoles and even small fish. They have jaw-like pincers on their head, which they use to grab their prey.

JAWS AND CLAWS

Down in the dark, murky world of bugs, there is a war going on! Even easy-going plant eaters need weapons to defend themselves. Mouthparts often have to do two jobs – killing and eating. Jaws and claws come in all shapes and sizes, and are best suited to each bug's lifestyle.

◀ Camel spiders can reach up to 15 centimetres in length. They have two powerful pincers attached to the front of their massive jaws.

Giant jaws

Camel spiders are equipped for battle. Their jaws are up to one-third of the size of their entire body. They use them to fight predators, or to grab prey, such as lizards. Then they pour ***digestive juices*** *over their prey before tucking in.*

WHAT IS IT?

ZOOM x20

Sticky claws and sucking straws

Assassin bugs use their powerful, sticky legs to grab hold of another insect. They inject a tube-shaped mouthpart into their prey, and pour digestive juices into it. This turns the prey's insides into a thick liquid, which the bug sucks up.

Creepy crawly

Centipedes are best known for having lots of legs, but they use one pair of feet as deadly claws. The front pair of legs has evolved into nasty, sharp claws that can inject deadly ***venom*** *into predators or prey.*

Mighty munchers

A plague of locusts can cause chaos. Their enormous ***mandibles*** *– the chewing and crushing mouthparts – are so strong and fast that a locust can eat its body weight in food in just one day. A swarm of locusts can be 1200 square kilometres in size – the size of a large city – and it can contain 80 million locusts. One swarm can wolf down 192 million kilograms of plants in one day.*

ANTS UNITED

The menacing face of an ant hides a brilliant brain. These little bugs have bigger brains than any other insect, with 250,000 brain cells each. Their real intelligence, however, comes from the way ants work together. Scientists think that a **colony** of ants operates like a super-brain. A colony can make decisions, working together in the same way that the cells inside our brains co-ordinate their efforts.

Driver ant

Hooked claws help the ant to climb over plants.

Ready for action

Ants have two main eyes, which can detect movement, and three smaller eyes, called ocelli, which are better at detecting levels of light. Their main sense, however, is smell. Strong jaws, or mandibles, can be used to tear and carry food, build nests or bite prey.

It is...

a queen herdsman ant, leading her colony to a new nest. From time to time, a queen decides to lead her whole colony nearer to food supplies, and a new nest is built. The colony may contain more than 10,000 workers and 4000 larvae and pupae.

FACTOID

Driver ants are the largest ants in the world. Working together, a colony attacks every animal in its path – including snakes and people.

VITAL STATISTICS

Common name	Driver ant
Latin name	*Dorylus nigricans*
Size	5–50 mm in length
Habitat	Rainforests and grasslands
Special feature	Can live in colonies of 20 million ants

Job share

Ants are called social insects because they live and work together. One queen lays all of the colony's eggs. Worker ants are wingless females (left) that do all the chores, including fetching food, nest-building and taking care of the eggs. Male ants grow wings and create swarms at mating time.

HIDDEN FROM VIEW

Even with the best camera equipment, or the sharpest eyesight, it can be almost impossible to find some bugs. Look closely at these pictures and be amazed at the many ways nature uses colour, pattern and shape to make an animal disappear from view.

◀ A **katydid**'s camouflage makes it blend into the tree bark, so it isn't seen by prey or predators.

Singing for your mate

When a bug is cunningly disguised as a leaf or patch of moss, how can it expect its mates to find it? The answer is to sing. Katydids are bush crickets that are masters of deception, but they alert mates to their presence with a loud and characteristic song that sounds like 'katy-she-did'.

WHAT IS IT?

Prickly problems
When treehoppers suck the tasty sap from inside a plant stem, they could be sitting targets for predators. They overcome this problem by disguising themselves as part of the plant. The young bugs look like the brown, knobbly bark, but the adults pretend they are green prickly thorns instead.

Pretty as a petal
Colourful crab spiders perch motionless on flowers, and calmly wait for lunch to walk by. These lurking predators are invisible because their colours are flawlessly matched to the colours of the petals.

Clever tricks
This trickery certainly works on birds, which normally wouldn't hesitate to swoop down and pluck a fat, juicy caterpillar from its feeding grounds. With a green body, the caterpillar is camouflaged among leaves, but that freaky face on the caterpillar's rear end is enough to make any sharp-eyed predator think twice.

BEAUTIFUL BUGS

Have you ever wondered how butterflies are able to produce such dramatic colours and patterns on their wings? The secret lies in the structure of tiny scales that are able to turn light into incredibly vivid blues, reds and greens. Some butterflies even appear to have a metallic sheen to their wings.

Creating colour

The overlapping scales on a butterfly's wing can have a range of colour ***pigments****, shiny reflective 'mirrors' and air spaces. These work together to absorb ultraviolet light, which our eyes can't see, and turn it into bright patches of blue and green. They work in a similar way to Light-emitting diodes (LEDs), which are used in televisions.*

Iridescent colours on a butterfly's wings change depending on the angle that you view them.

It is...

an owl butterfly, flaunting its eerie-looking eye-spot. The flash of the 'eye', when the butterfly flutters its wings, is startling, and might make a predator think twice before attacking – all the time the butterfly needs to make its escape.

Peacock butterfly

FACTOID

The peacock butterfly's wings make a strange 'hissing' sound when they open to scare away predators.

VITAL STATISTICS

Common name	Peacock butterfly
Latin name	*Inachis io*
Size	50–70 mm (wingspan)
Habitat	Gardens and meadows
Special feature	Four false eyes on wings

Short but sweet

The larvae of butterflies and moths are called caterpillars, and their job is to feed and grow. This caterpillar's stunning skin is a colourful warning that it is covered with stinging bristles. After metamorphosis this impressive little beast will emerge from a cocoon as a drab brown moth.

ON THE MOVE

Bugs are amazing. Their hugely adaptable bodies have evolved the ability to move in all types of ways. There are bugs that can fly, ones that crawl and others that can swim. And there are even some that can do all three!

◀ Being able to fly is an essential life skill for an emerald euphoria beetle. It moves from flower to flower, eating nectar and pollen.

Gripping force

Spiders are able to crawl over smooth surfaces because they have tufts of tiny hairs under their claws. Each hair is further divided into thousands of tiny 'end feet', which help the spider to stick to the smoothest of surfaces, including glass.

Elastic jumpers

Tiny cat fleas can jump up to 34 centimetres in a single leap, and they can keep jumping for several days without stopping for a rest. Fleas have ultra-elastic material, called resilin, in their legs, which can be squeezed and released like a spring.

Fast bugs

Tiger beetles are among the fastest of all bugs. The colourful Australian tiger beetle can chase its insect prey at 9 kilometres an hour, before tearing it apart with large mandibles.

Swimming and diving

Water boatmen are able to swim and dive, thanks to their final pair of legs, which are shaped like oars. They are long, and fringed with hairs. These bugs can survive underwater because they carry bubbles of air next to their body.

WHAT is it?

HONEY MONSTERS

Most of us think twice before getting up close and personal to a bee. That buzzing sound, those warning stripes and the fear of being stung are all good ways to keep us – and predators – at arms' length. In fact, bees are among the most important bugs on the planet. Without them, flowers would not grow seeds, and we would quickly run out of food.

The chest area, or thorax, is packed with muscles used for flying.

FACTOID

Bees travel 90,000 kilometres, and visit more than two million flowers, to make a single jar of honey. One honeybee will produce just half a teaspoon of honey in its whole life.

VITAL STATISTICS

Common name	Honey bee
Latin name	*Apis mellifera*
Size	0.5-2.5 cm in length
Habitat	Anywhere that flowers grow
Special feature	Are able to turn nectar and pollen from flowers into sweet honey

Super storage
Bees have pollen baskets on their legs, which they use to store pollen that they collect from flowers. Worker bees use the pollen to make honey, and to feed the colony.

Home sweet home

A beehive is home to the colony. The queen lays eggs in wax cells, where they are tended by workers. When the eggs hatch, workers feed the larvae. Most of the larvae will grow into workers. Some of them will grow into males, which are called drones. A few will grow into queens.

It is...

an orchid bee. These are the only animals, apart from humans, that are known to make perfumes. Males collect scents from orchid flowers and mix them with other ingredients, such as fruit, to make a perfume that attracts females.

SPOTS AND STRIPES

Patterns, spots and stripes decorate many bug bodies. Certain patterns may help a bug to remain hidden in the undergrowth, or they may make them more attractive to mates. Bold patterns and colours are often used to warn predators that the bug tastes foul, or stings.

Dressed to impress
Glamorous male ladybird spiders use their red back, black spots and white-banded legs to impress females, who are dark and dull by comparison. Each female can only lay one ***brood*** *of eggs because when they hatch, the little* ***spiderlings*** *gang up and eat her!*

Pretty in peach
Velvet worms have delicate colours, and often patterns on their soft, squidgy body. These animals live in dark, moist habitats and can shoot a jet of slime to catch their prey.

Spiral puzzle

Grove snails with light-coloured shells live in warmer places than those with dark shells. No one knows why some grove snails have dark spiral bands, while others have plain shells.

Blue genes

How do bugs create colours and patterns on their bodies? Scientists think it might be the work of a special ***protein****, called morphogen. When morphogen reaches certain parts of the bug's body, it tells them to start making pigment, which creates colour. It is the bug's* ***genes*** *that decide where the morphogen will have this effect.*

WHAT is it?

HAIRY NOT SCARY

Imagine you could shrink to the size of a bird-eating spider, and come face to face with one of these mini-beasts. These are the largest of all spiders, and some of the biggest bugs on the planet. Thankfully, bird-eating spiders – also known as tarantulas – look more deadly than they actually are.

A bad rep

Lots of people are scared of spiders, but they don't deserve their bad reputation. Very few spiders are capable of hurting humans, and spider bites are very rare events. In fact, spiders are incredibly important. They kill flies and other disease-bearing bugs, and are food for billions of other animals, including mammals, birds, reptiles and amphibians.

Touchy-feely

Spiders have little hairs all over their body. The hairs are essential to a spider's survival. They are extremely sensitive to touch and vibration, so they alert the spider to the presence of another animal nearby. Even blind spiders, such as those that live in caves, can find and catch a fly, just by using this type of information.

FACTOID

Spiders do not have a tongue or a nose. They can smell and taste things by using the special hairs on their legs.

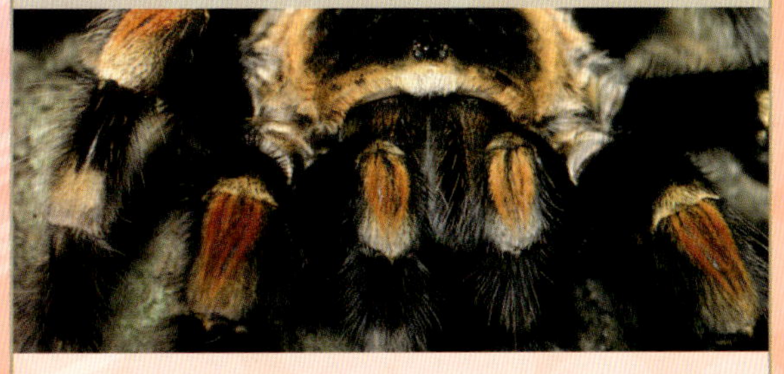

VITAL STATISTICS

Common name	Mexican redknee tarantula
Latin name	*Brachypelma smithi*
Size	13–18 cm in length
Habitat	Scrubland and deserts
Special feature	Can live for more than 20 years

It is...

a female wasp spider, with bright warning stripes. Males are small, brown and easy to miss. These spiders are orb-weavers, which means they create complex webs of silk. They use the webs to catch their prey, which are killed with a venomous bite.

USE YOUR EYES

Study these zooms that appear throughout the book. Can you recognize any of them just by looking at them? Are there any clues, such as colour, bodyparts or shape, that help you work out where you've seen these images before?

1 *As I flutter by you will see flashes of vibrant colour.*

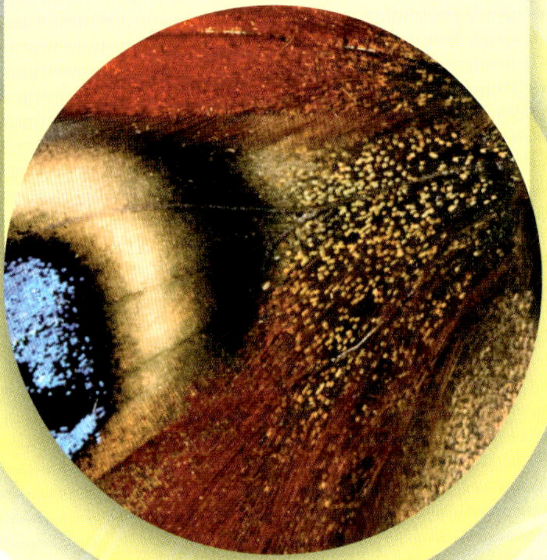

2 *I pray for my supper, but beware my striking pose. Who am I?*

3 *I am a giant Scolopendran, but you know me better by my common name.*

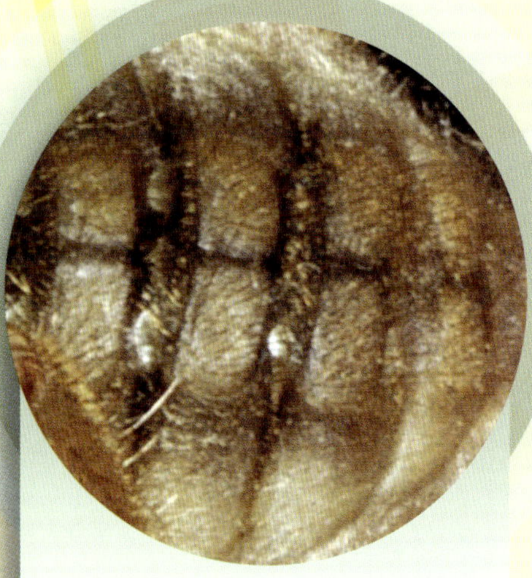

4 *I am a desert fighter with mighty mandibles.*

5 *I am the sweetest of bugs with a sting in the tail.*

6 *Cats might think I am an irritating pest.*

7 *I spend all of my time eating and growing. When I grow up, I want to be a giant beetle, with legs and wings.*

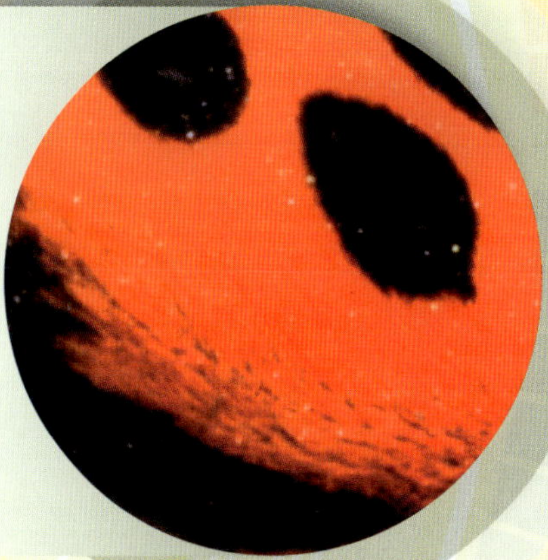

8 *My name could fool you. I'm no lady and I have eight legs, not six (or even two!).*

9 *Who are you looking at? This giant eye is a clever trick.*

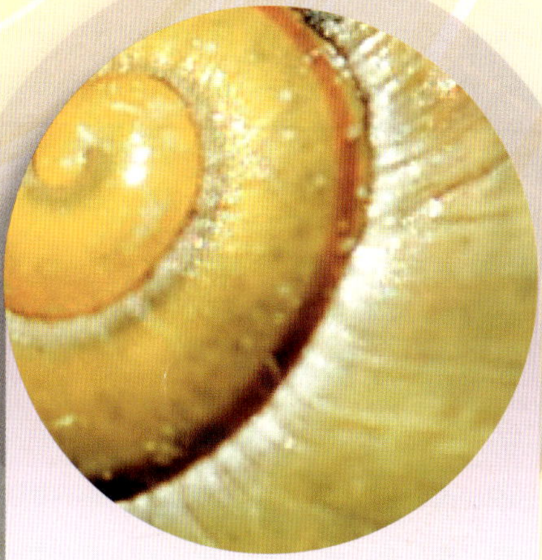

10 *You could get dizzy following my lovely spiral.*

11 *My eyes are bigger than my brain but are you smart enough to name me?*

Answers: 1. p18 2. p6 3. p13 4. p12 5. p22
6. p21 7. p9 8. p24 9. p17 10. p25 11. p10

GLOSSARY

Brood All of the eggs, or young, that are produced at one time.

Camouflage Colours or patterns that help an animal to stay hidden from view.

Cocoon The silken case spun by an insect to protect the pupa.

Colony A group of animals that live closely together.

Cuticle The tough outer skin of a bug.

Digestive juice When an animal digests food, it breaks it down into smaller parts that can be absorbed into its body. Its body makes powerful liquids, called digestive juices or enzymes, to do this.

Eye-spot Some animals have patterns on their body that resemble the eyes of larger animals. These eye-spots may help to scare away predators.

Gene Animal cells contain genes, which hold all of the information necessary for the animal to live, grow and reproduce.

Grub Young, soft-bodied bugs are larvae, but they are also sometimes called grubs or maggots.

Iridescence Colours that seem to change when they are seen from different angles are described as iridescent.

Katydid A green or brown insect that belongs to the same family as crickets and grasshoppers.

Larva (plural: larvae) A young, soft-bodied bug. It will eventually change and grow into an adult.

Lens A transparent object with curved sides that gathers light and bends light rays. Lenses in eyes focus light rays into the back of the eye, so an animal can see. Lenses in cameras and microscopes are made of glass and can magnify an image.

Macro lens Used in a camera to take photographs close to the subject, such as a bug.

Mandible The mouthparts of a bug.

Mantid This insect is also known as a mantis, and is related to the cockroach.

Metamorphosis The way that a young bug, or larva, changes into an adult.

Moult When a bug grows bigger and sheds its old skin, or cuticle.

Pigment Chemicals that give an animal colour.

Proboscis A long, slender mouthpart that some bugs absorb liquid through.

Protein Essential substances that all living things need to grow, because cells are made of protein.

Pupa The stage of an insect's lifecycle when it is going through the change from larva to adult. It is in a tough case, called a chrysalis, which protects it at this time.

Spiderling A young spider that has just hatched from its egg.

Ultraviolet A type of light that cannot be seen by humans, but is visible to many other animals – especially insects.

Venom A type of poison that is made by animals and injected into another animal's body, often by biting or stinging.

INDEX

NOTES FOR PARENTS AND TEACHERS

Photography and microscopy are two ways that the physics of light and lenses can be applied to our everyday lives. Use the Internet* to find diagrams that show how lenses bend (refract) light that goes through them. Look at diagrams that show both convex and concave lenses, to discover how the shape of the lens changes the effect. Together, you can work out which of these two types of lens is used in microscopes, telescopes and binoculars. You can also use the Internet to explore the role of lenses in the human eye, and how corrective lenses in spectacles improve eyesight.

On a sunny day, you can demonstrate the focusing power of a lens. Hold a magnifying lens just above a piece of paper that is laid out in sunshine. Angle the lens until the light is focused on the paper as a small bright dot. As it heats, the paper will smoke and burn.

It is easy to make a water lens that shows how even a simple lens can magnify images. Lay a piece of clingfilm or other transparent plastic over a piece of newspaper text. Use a syringe or a teaspoon to place a single drop of water on the plastic. You will notice that the text beneath the water drop is magnified. Find out what happens when you make the drop bigger, or smaller.

Go back through the book and choose certain bugs to investigate together. Encourage the child to find out how big each animal is, how it lives its life and where in the world they might encounter it. Children can zoom in on wildlife for themselves. A simple hand lens will enable a child to see bugs in close-up. Other useful equipment includes a camera, a sketch pad, a pencil and a ruler (for taking measurements). Many naturalists discover a love of wildlife as children, simply by observing animals, sketching and photographing them, and taking notes about their behaviour.

Teach children to respect the wildlife around them. They can watch wildlife without harming it. Encourage them to respect the animals' habitats and to disturb the environment as little as possible. Remind them that some bugs and plants can sting, so they should exercise care.

*The publishers cannot accept any liability for the content of Internet websites, including third-party websites.